THE ESSENE SCIENCE OF LIFE

This book is dedicated to the nurses and patients of the Leper Colony at Orofara, Tahiti, who in 1933 under my supervision so successfully applied these methods. With faith and courage, they re-enacted THE ESSENE GOSPEL OF PEACE, proving that the Essene Way of curing leprosy is as valid in the twentieth century as it was two thousand years ago.

THE ESSENE SCIENCE OF LIFE

as applied to Health and Disease
by the First Century Essene Brotherhood at the Dead Sea
and translated into contemporary terms based on the

ESSENE GOSPEL OF PEACE

The Aramaic and Old Slavonic Texts
compared, edited and translated by

EDMOND BORDEAUX SZEKELY

MCMLXXXVI
INTERNATIONAL BIOGENIC SOCIETY

This Anniversary Printing is Limited to 2000 Copies.
It Contains the Author's Own Revisions of 1978 and Illustrations from the Archives of the International Biogenic Society which Appear for the First Time in Publication.

Cover: The Essene Science of Life in action in the form of sun, air and water therapies at Rancho La Puerta, Tecate, Baja California, Mexico, some time in the early forties (see *Search for the Ageless, Volume Two: The Great Experiment,* by Edmond Bordeaux Szekely.)

Book Design by Golondrina Graphics
Archival Photographic Reproduction by Centaur Photography

EDMOND BORDEAUX SZEKELY
(circa 1935)

"Only thus can we obey the principle of practical altruism: to dispel evil not by attacking it but by maintaining and strengthening the good. To overcome ignorance with education, egoism with altruism, violence with constructive social activity, misery by the removal of its causes, disease by the strengthening of the organism's vitality, and chaos by the organization of order. To create a better future, not by the destruction of the past, but by the constructive and progressive amelioration of the future."

—page 61, *Ibid.*

SOME BOOKS BY EDMOND BORDEAUX SZEKELY

THE ESSENE WAY—BIOGENIC LIVING
THE ESSENE GOSPEL OF PEACE, BOOK ONE
BOOK TWO, THE UNKNOWN BOOKS OF THE ESSENES
BOOK THREE, LOST SCROLLS OF THE ESSENE BROTHERHOOD
BOOK FOUR, THE TEACHINGS OF THE ELECT
THE DISCOVERY OF THE ESSENE GOSPEL: The Essenes & the Vatican
SEARCH FOR THE AGELESS, in Three Volumes
THE ESSENE BOOK OF CREATION
THE ESSENE JESUS
THE ESSENE BOOK OF ASHA
THE ZEND AVESTA OF ZARATHUSTRA
ARCHEOSOPHY, A NEW SCIENCE
THE ESSENE ORIGINS OF CHRISTIANITY
TEACHINGS OF THE ESSENES FROM ENOCH TO THE DEAD SEA SCROLLS
THE ESSENES, BY JOSEPHUS AND HIS CONTEMPORARIES
THE ESSENE TEACHINGS OF ZARATHUSTRA
THE ESSENE SCIENCE OF LIFE
THE ESSENE CODE OF LIFE
THE ESSENE SCIENCE OF FASTING AND THE ART OF SOBRIETY
ESSENE COMMUNIONS WITH THE INFINITE
THE FIRST ESSENE
THE BIOGENIC REVOLUTION
THE ORIGIN OF LIFE
THE COSMOTHERAPY OF THE ESSENES
THE LIVING BUDDHA
MAN IN THE COSMIC OCEAN
TOWARD THE CONQUEST OF THE INNER COSMOS
FATHER, GIVE US ANOTHER CHANCE
THE ECOLOGICAL HEALTH GARDEN, THE BOOK OF SURVIVAL
THE TENDER TOUCH: BIOGENIC FULFILLMENT
THE DIALECTICAL METHOD OF THINKING
THE EVOLUTION OF HUMAN THOUGHT
THE GREATNESS IN THE SMALLNESS
THE SOUL OF ANCIENT MEXICO
THE NEW FIRE
ANCIENT AMERICA: PARADISE LOST
PILGRIM OF THE HIMALAYAS
MESSENGERS FROM ANCIENT CIVILIZATIONS
SEXUAL HARMONY: THE NEW EUGENICS
LUDWIG VAN BEETHOVEN, PROMETHEUS OF THE MODERN WORLD
THE FIERY CHARIOTS
CREATIVE WORK: KARMA YOGA
THE ART OF STUDY: THE SORBONNE METHOD
COSMOS, MAN AND SOCIETY
THE BOOK OF LIVING FOODS
SCIENTIFIC VEGETARIANISM
THE CONQUEST OF DEATH

Book Design by Golondrina Graphics

THE ESSENE SCIENCE OF LIFE

Contents

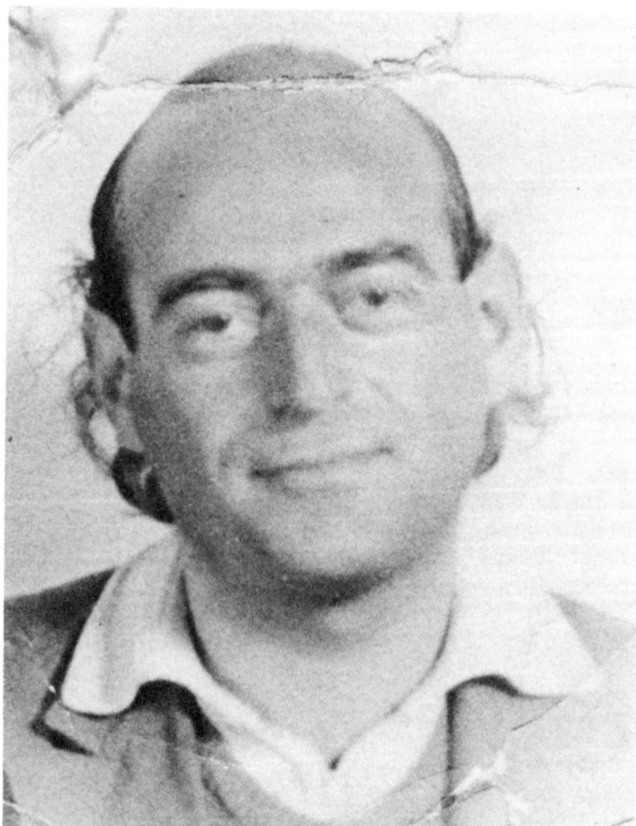

Edmond Bordeaux Szekely *(circa 1945)*

PREFACE

Most of the lectures which comprise this book were published in 1937 by the participants of an unforgettable Seminar which took place in 1935 in a peaceful mahogany forest on the bank of the Rio Corona River in Tamaulipas, Mexico. The story of this Seminar is told in the first volume of my memoirs, *Search for the Ageless: My Unusual Adventures on Five Continents,* and excerpts from the Seminar material collected by the participants, a gifted band of rebels from the healing profession (young medical doctors, osteopaths, chiropractors and naturopaths), are contained in another of my books, *The Cosmotherapy of the Essenes.* The content of this book, as well as the above-mentioned material, were both included in a rather voluminous book of 700 pages with the title, *Cosmotherapy, the Medicine of the Future,* out-of-print for more than forty years. The Introduction to the book survives, however, in the form of this booklet, and now I have tried to bring it up-to-date, a task which has been somewhat depressing, as it consisted mainly of including phenomena less well-known in the thirties, such as chemicals in foods and pollutants in the air.

Man is the quintessence of nature, and has all its elements within him. Its forces are constantly acting upon him, but he may react and make them subservient to his will. Observing the relations of cause and effect, he is King; but failing in this, they become his masters, inflicting as penalties all manner of disease.

As the correct use of the laws of nature known to science would prevent most cases of sickness, and annually save millions of lives, how important it is that people be taught the best methods of utilizing the natural forces and laws!

Education is the only remedy. Let all be informed upon the basic biological functions of the organism and the prevention of disease.

In this booklet, plain, simple language has been used and technical terms avoided. Though I have updated the information, the laws of nature have not changed, and the simple wisdom here given will prove most valuable in preventing sickness and in enabling all to cooperate better with Nature in her best efforts to save life.

Through understanding of the natural laws and the utilization of the natural forces, with perseverance you will progressively acquire Health, Youth and Longevity.

This was the purpose of that long-ago Seminar in the mahogany forest in Mexico. Many years have passed since then, but the quality of dedication and perseverance on the part of a sincere seeker of truth has not changed, nor will it ever change. I invite that seeker of truth to read carefully the pages of this booklet, which has helped tens of thousands over many years to start a new life of health and happiness.

EDMOND BORDEAUX SZEKELY

Mission, British Columbia, August 10th, 1978.

Edmond Bordeaux Szekely
(circa 1930)

INTRODUCTION

HOW TO BEGIN TO FIND HEALTH

By health we mean not merely freedom from pain, discomfort and disease. We mean possession and exercise of vitality. There should be conviction, energy, purpose, in the life of every individual. It is not good enough to just get along from day to day, constantly worried and ill at ease, perhaps seeking relief from despondency in over-excitement which only makes the daily round seem all the more depressing. Health that means joy of living and working should be, and can be, the possession of every individual. Make up your mind that you want it and are determined to get it. That is the first and essential step. Then realize that the way to health depends upon your making use of the knowledge of natural laws and forces. We cannot make you healthy, *you* are the only person who can do that. We can help you by telling you how. You will require patience, persistence, strength of will, and these you can build up if you do not already possess them.

THE CAUSE OF DISEASE

You must be prepared to shed many conventional ideas and to re-examine many habits and customs which you have taken for granted. But we shall never ask you to accept any idea without scientific reason for so doing, for our work is based upon knowledge and experience, and all our theory is proved by practice and results.

Let us first explain how we regard the problem of health. We know that all disease, whatever its name, is due to one fundamental cause—the presence in the body of impure and unsuitable matter. Such matter, because it cannot be absorbed and used by the body, acts as a poison and begins to injure the blood cells and various organs. Thus, disease begins. Such poisonous and harmful matter are called waste matter.

The body tries, without our being conscious of its efforts, to expel such harmful matter. Thus symptoms appear, sometimes accompanied by pain. When the presence of such harmful matter has reached a dangerous proportion a crisis develops, the body has perhaps a fever, or some other visible and violent sign of disorder. If the fever abates, it means that the dangerous excess of poison has been thrown off, the patient "recovers." The process of getting rid of toxic matter we call "elimination." But the process by which the harmful matter is being accumulated continues if the patient goes back to his old mode of living. Since the symptoms of disease are the effect and not the cause, treatment which aims only at removing or suppressing the symptoms, obviously cannot remove the cause. A favorite idea is that "germs" cause disease. By germs is meant a very small form of living organism which is parasitic in the body. Certainly germs exist, many are very harmful, and such kinds are undoubtedly present in large quantities in various forms of disease. Enough of them may bring about death.

But the question is, why are germs able to multiply to a dangerous extent in some bodies and not in others? We say that it is the preliminary presence in the body of poisonous matter caused by improper eating, weakening it, that enables the germs to gain their hold.

In other words, germs and parasites of all kinds are rather a secondary accompaniment of disease, not the basic cause. Everyone knows that a healthy body is able to deal with germs and live, when a weakened body will succumb.

HOW TO AVOID AND OVERCOME DISEASE

We are not going deeply here into the complications of this question. What we aim to do is to give you our conception of health and disease in order that you may carry on the various recommended exercises and practices intelligently and with understanding, and not merely mechanically.

If all disease has its beginning in the presence of harmful matter in the body, it follows that the way to health is to keep all such substances out of the body and to introduce into it only those substances which it requires and which are beneficial to it. We shall need also to introduce such beneficial substances only in the quantities which are needed, and at the times when they can be absorbed.

Since the body in the course of its normal and healthy activity converts nutriment and air into energy and movement of various kinds, we must also consider the chemical and other changes which go on inside. These produce what we might call by-products and waste material which will also be poisonous if they are not eliminated.

Therefore, we see that the process required is a double one. We must supply the body with all its needs of helpful nutriment and see that it rids itself of all waste and by-product material.

BODY AND MIND

When we speak of the body, we must not forget that the brain is a part of it. Mental activity has just as much of a physical basis as has purely muscular action, like kicking a football.

What we think also affects our physical condition. So we need to take as much care of our mental nutriment as we do with our food.

In fact, we find that our whole organism, the sum total of our physical and mental organs and faculties is essentially a unity. An injury to one part will injure the whole system. So our effort to achieve health must take the form of an all-round effort and not be limited to any one area, even if we have distressing symptoms of pain or disease apparently confined to one area.

We must also not forget that the body is made for work. We do not want to spend all our time in being healthy as an aim in itself. We want to achieve health so as to be able to carry on with energy and joy and efficiency our work in the world, whether we are craftsmen, mothers, artists, athletes, laborers or clerks. Whatever form of creative work* we have

*Recommended reading: *Creative Work: Karma Yoga,* by Edmond Bordeaux Szekely, available from the International Biogenic Society.

an affinity with, health enables us to give our best, and we should seek it for that purpose.

What we have said now explains to you the background for the practical advice we are now about to give you, and which if you carry it out will prepare the way toward Health, Youth and Longevity.

In setting out our advice we have done so taking in mind the fact that most people have to carry on their daily work whether they are ill or not and consequently cannot give themselves holidays, a stay in the country, and such things. If you are one of those who has such possibilities then you will get much more rapid results.

THE BASIS OF COSMOTHERAPY

THE BIODYNAMIC UNITY OF MAN, COSMOS AND THERAPIES

Man	*Cosmos*	*Therapies*
Bones	Earth	Geotherapy (mineral therapy, clay and mud applications)
Blood	Water	Hydrotherapy (external: water baths; internal: fresh juices)
Skin and Lungs	Air	Aerotherapy (air baths, rhythmic breathing)
Nerves	Sun	Heliotherapy (sun baths)
Muscles	Vegetation	Nutrotherapy (diet)
Movement	Movement	Dynamotherapy (physical culture, exercise)
Cell	Atom	Cellulotherapy (fasting)

THE BASIC BIOLOGICAL FUNCTIONS OF THE ORGANISM

The bones in the human organism represent the solid element in the same way that the lithosphere (earth) represents it in nature; besides, our bones are constructed with the inorganic minerals of the earth which enter our organism in organic combinations through the assimilation of fruits and vegetables.

The circulation of the blood in the human organism has the same role and function as the circulation of water in nature: it cleanses, dissolves, eliminates.

Respiration by the skin and lungs has the same role and function in the organism as the circulation of air in nature. It refreshes and everywhere distributes the oxygen necessary for life.

The nerves have the same role and function in the organism as the solar rays in nature; without the sun's rays there would be no terrestrial nature; without the nerve vibrations the human organism would die.

The muscles in the organism unite the bones, blood, skin, lungs, and nerves, just as the vegetation in nature unites the earth, water, air, and sunshine.

Movement has the same role in the human organism as the atom is in nature; they both have the same function—they sustain the reciprocal transformation of the various energies.

THE BASIC PRINCIPLES OF BIOGENIC COSMOTHERAPY

There is a single disease: inharmony, whose symptoms are variously manifested, depending on the extent and mode in which natural law has been violated. This is the law of disease.

There is a single health: harmony, which is manifested in permanent evolution towards perfection, in proportion as we are able to create and utilize the preconditions of existence in harmony with the laws of nature.

Illnesses are produced by struggle between the natural vitality of the human organism and inferior, degenerating processes.

The symptoms of disease are without exception the result of this battle going on in our organism.

Man is a microcosm in the macrocosm. He lives in a field of permanently varying forces. His health is harmony with the surrounding cosmic and natural forces; his disease, physical or mental, is disharmony with them. Cosmotherapy is the science of re-establishing the disturbed harmony between man and the natural and cosmic forces. It is at once the most ancient and most modern system of healing, practised by the greatest Masters of all ages: Zarathustra and the ancient Sumerians, Buddha, Patanjali and the Sixteen Yogas, Confucius, Lao Tzu, the Egyptian and Greek Therapeutae, the Pythagoreans, Hippocrates, the Essenes, Medieval and Modern Natural Therapeutics, etc., constantly verified by modern natural research and discovery.

THE BASIC BIOLOGICAL NEEDS OF THE ORGANISM

THE SEVEN PILLARS OF COSMOTHERAPY

Only those natural and cosmic forces are capable of healing the organism which constantly sustain it.

I. The largest surface in the organism is the skin with the lungs. For this reason the most fundamental need of the organism is air. The organism consumes approximately seven hundred quarts of air an hour.

II. The second fundamental need of the organism is water, for the greatest part of the organism itself consists of water.

III. The third fundamental need of the organism is the rays of the sun, for without light the human nerve system is unable to function, and the activity of the organism is directed by the nerves.

IV. The fourth fundamental need of the organism is food, for it accumulates the air, water and sun in concentrated nutritive form capable of constructing and nourishing the muscles and of giving them the necessary energies for work. In addition, foods construct and nourish the cells of the skin and lungs, blood, nerves, and bones.

V. The fifth fundamental need of the organism is movement, for in the universe, too, everything moves; there is not a single fixed point.

VI. Just as the universe constantly needs to change its atoms, so does the human organism constantly need to change its cells. In the universe, the atoms are dissociated and created by various concentrations of cosmic radiations; in the human organism the cells are dissociated by the various concentrations of nerve energies (fasting, processes of elimination, etc.). This is the sixth fundamental need of the organism.

VII. The organism always needs and uses the same substances as those of which it is constructed. Similarly, earthly life only uses those chemical elements which our planet, the Earth, contains.

THE BASIC METABOLISM OF THE ORGANISM

Hence comes the most fundamental law of cosmotherapy: an organism constructed of unhealthy substances always desires the same unhealthy substances up to the moment when all its unhealthy cells have been replaced by healthy ones. For instance, an organism whose cells are constructed (saturated) with meat, alcohol, nicotine, etc., always desires meat, alcohol and nicotine. Similarly an organism whose cells are constructed of fruits and vegetables always desires fruits and vegetables. Finally, an organism unaccustomed to water, air, and sun bathing, in the beginning has an antipathy to these things, but once it becomes accustomed to them it desires them constantly. That is why the first half of the cosmotherapeutic treatment (up to when half of the cells of the organism are renewed) is difficult and demands much strength of will, as the old cells of the organism constantly clamor for the old food and fight against the new and unaccustomed substances.

In consequence, the only way in which you can succeed in renewing your cells, rejuvenating yourself and prolonging your life considerably, is by concentrating your intelligence on the understanding of the cosmotherapeutic laws given in the following chapters and by educating your will to overcome your unhealthy desires and the disagreeable symptoms

Professor Dr. Edmond Bordeaux Szekely *(standing in river)* with a group of students of the Essene Science of Life in Tahiti, 1933. The story of his remarkable success in introducing the Cosmotherapeutic method of treatment of leprosy at the Orofara Lepers Colony is told in the first volume of his memoirs, *Search for the Ageless, Volume One: My Unusual Adventures on Five Continents.*

of elimination and of cell-changing and, in addition, by resisting the influence of ignorant talk and criticisms of unbalanced individuals around you.

If you have made up your mind to carry out a complete reform of your organism and your life, then study the following chapters where, in abbreviated form, you will find the fundamental laws of Cosmotherapy.

WATER, SUN, AIR, MOVEMENT

BATHING AND THE WATER OF LIFE

Bathing includes water, sun and air bathing.

Bathing removes impurities from the surface of the body and also stimulates the skin and circulation of the blood.

The best bath is one in which the water is vigorously moving, the aerated activity producing a hydromassage far superior to any other kind. The modern "whirlpool" bath, provided it is not chlorinated, best approximates bathing in an outdoor, fast-running stream. The shower is the best form of indoor bath, much superior to the ordinary bath. If there is no shower installation, any bath tub can be fitted with a rubber pipe and shower nozzle.

A daily bath is an absolute essential to health. The bath should follow the exercises immediately on rising, and should be a regular practice. A bath before retiring is also highly valuable.

Cold water is the most satisfactory bath. But if you are not used to it, begin with tepid and gradually get used to cold. Do not begin with a heroic effort to stand the shock of a cold immersion, you will only injure yourself. But accustom yourself gradually to the cold water and you harden the skin and the body. Particularly apply the cold water to the back of the neck and head and to the chest.

Baths in hot water are beneficial, but not so valuable as cold baths. If the occupation is such that frequent warm washing with soap is necessary to remove grime, sweat, etc., then the bath should always be concluded with a cold shower.

SUNBATHING AND THE FIRE OF LIFE

Without sun, no life on earth. The limits of the zones of

life are the limits of the zones of the sun's rays (alike on the surface of the earth, in the oceans and in the atmosphere). Without sun there would not be the right temperature for life, neither for plants nor animals nor men. The source of life is the combination of air, water, and sun.

The best and most natural source of Vitamin D for the human organism is sunbathing.

The rays of the sun have the multiple benefit of strengthening, disinfecting, healing and eliminating.

PRACTICAL INSTRUCTIONS

When taking a sunbath, always lie with the head away from the sun, so as to have the top of the skull in the shade. Lie for ten minutes on your back and for ten minutes on your stomach. Ideally, choose a sandy or grassy spot beside the water and if the ground is covered with pebbles lie on a mat.

TIME AND DURATION

It is very important that sunbathing be done with moderation and at the proper time, preferably in the morning hours when the sun is not directly overhead. If a sunbath is taken at midday, or for longer than about 20 minutes, there may be an increase of blood pressure, increase of pulse rate, sun burn or precancerous spots on the skin. Also, indiscriminate exposure to the sun is dangerous when the organism is intoxicated, as these toxic substances ferment under the influence of the thermic and chemical action of the sun and are thus able to cause serious crises in the organism.

SUNBURN AND ITS REMEDIES

If the path of moderation is followed, the skin will acquire a *slight,* healthy sun tan. A sun burn can be as serious as any other kind of burn and should be prevented. However, if it happens that enthusiasm makes one neglect precautions, mild sunburns can be cured without difficulty by covering the inflamed part with a little oil mixed with one-fifth part of lemon juice. Care must be taken during the daytime to leave the skin uncovered in shade and exposed to the air.

During the night put a double damp water compress, freshly prepared, over the burn and wrap some dry cloths around the compress. Change the compress once or twice in the course of the night. As long as the burn remains painful, that is, for two or three days, avoid exposure to the sun, but do not neglect the baths or the exercises.

THE BASIC RULES OF SUNBATHING

1. Do not lie down with your head facing in the direction of the sun.
2. Do not cover the skin with heavy oil, which only will stop up the pores. Massage with just a few drops of oil beforehand and oil your thirsty skin when you are out of the sun.
3. Take short and frequent sunbaths rather than long ones. Short baths are relaxing and advantageous, whereas long ones create nervous tension. Furthermore, the deep tan which they engender is only a defensive measure of nature that prevents the skin from drawing full benefit from the sun's rays. Excessive exposure also accelerates aging of the skin.
4. Start with five-minute sunbaths, gradually increase them to 20 minutes (ten minutes on each side) and try to take a 20-minute sunbath once a day.
5. If you have high blood pressure, or heart or lung trouble, be extremely cautious with sunbathing until the new cosmotherapeutic regime has normalized the organism in these areas.

HOW TO EXERCISE AND BATHE IN THE CITY
WHAT TO DO DURING WINTER

In cold seasons and climates and in large towns where it is difficult to find outdoor water, sun and air, the open air baths and exercises may be replaced with swimming in an indoor pool. The water should have an agreeable temperature and the air should also be fresh (well ventilated).

BREATHING AND THE AIR OF LIFE

Breathing is our first and most important function. Breath is the link between the individual and the universe. Where there is breath there is life. Where there is life there is breath.

Do simply deep breathing with the window open. Begin when you rise and work up to ten minutes. Breathing means exhaling to the fullest extent, remaining with the lungs empty for a few seconds, inhaling slowly to the fullest extent, retaining the breath for a few seconds. When you remember that chemical processes equal to decomposition take place at each breath, you will see the importance of completely emptying the lungs after each breath.

It is never necessary to strain or to labor over breathing exercises. If you feel tired after a few breaths, then stop, but the next day do one or two more, so you will improve.

During the day, when walking or out in the air, repeat these exercises often. Before you sleep, breathe deeply for five minutes and your rest will be more beneficial.

Air is our most important food, and it goes without saying that if there is anything you can do to assure that the air you are breathing is clean and unpolluted, by all means do it.

EXERCISE, THE BUILDER OF STRENGTH

The importance of exercise is that it improves the circulation of the blood and tones up all the organs, thus greatly aiding in the elimination of harmful matter from the system.

We are not speaking of exercises that require special apparatus and equipment, nor of training for particular forms of sports or athletics. We are speaking of the minimum daily exercise required and possible for everyone no matter what their circumstances, to keep the body fit.

After breathing, the most important physical requirement is exercise: to stimulate the circulation and to move all the muscles of the body. Modern life usually means a sedentary way of living, and most people have many muscles which are never moved at all. This allows all sorts of waste matter to collect in different parts of the organism.

Here are a few easy exercises, starting with two which should be done every morning. Remember the important thing is to work up gradually to an improved level.

Stand on a rug with as little clothing as possible. Open the windows.

DAILY EXERCISE NO. 1: TRUNK TURNING AND FORWARD BENDING

Stand with the feet placed parallel with each other, that is, with the heels the same distance apart as the toes, about seven inches from each other. Stretch out the arms to either side, level with the shoulders, at the same time clenching the fists. Keep the arms in that position while making an energetic twisting movement of the trunk from the hips towards the left. Next do a forward bend stretching the right fist as low as possible, while the left fist is lifted up behind to a vertical position. Stand straight once more, turn the trunk—this time to the right—bend forward again, stretching the left fist down and the right fist upwards. The thighs are always kept taut. As time goes on, this exercise can be repeated a great many times.

DAILY EXERCISE NO. 2: LEG MOVEMENTS

Lie on the back, slightly raising the nape of the neck. Keeping the hands beside the body, raise the legs and stretch them vertically, then separate and lower them without letting the heels touch the ground. Then start again and repeat the movement more and more frequently as training increases.

WEEKLY EXERCISES

Two excellent exercises which should be done weekly are (1) swimming with rhythmic breathing, a collaboration of water, air and sun, and (2) running with rhythmic breathing, a collaboration of earth, air and sun.

These two vitalizing exercises improve the basic biological functions of the organism. For this reason we advise you to practice these basic exercises *at least* once a week, according to individual possibilities. You must begin with a little at

Professor demonstrates Trunk Turning and Forward Bending at a Seminar at Lake Elsinore, California, around 1936.

first and then slowly and progressively increase them, always alternating the exercises with rest.

Swimming is the best sport for health, particularly in fresh water and the open air and sunshine. During the winter and when outdoor swimming is not available, a weekly visit to an indoor pool is a valuable investment.

INTESTINAL HYGIENE DURING THE COSMOTHERAPEUTIC TREATMENT

PREPARATION OF THE ORGANISM

At the beginning, in order to prepare the organism for the taking of an enema, one should eat fresh fruit juices for a whole day and no other food. Later on fasting for half a day will be sufficient to prepare the organism.

HOW TO TAKE AN ENEMA

About a pint to a quart of water should be absorbed without forcing the intestines. One should retain the water as long as possible without exertion. One should try a bowel elimination before taking the enema, so that the intestines may get rid naturally of all the superficial fecal matter or gas.

METHOD OF ADMINISTRATION

1. The container for the water should be fixed three feet high so as to provide the minimum pressure required to make the water flow into the intestines. 2. If the container does not hold enough water it must be refilled. 3. Put some vaseline on the nozzle at the end of the tube and mix together hot and cold water at the temperature of the human body. 4. Next, let a little water flow from the tube to get rid of the air in it. Then one should lie down on the left side, put the nozzle slowly in position oneself with the right hand and then open the tap above the nozzle. 5. When the water has penetrated, stay on the left side for five minutes, and then rest five minutes on the stomach, right side and back, in that order. In the latter position massage the stomach by rubbing it with the hand in clockwise direction. Stand up at the end of five minutes and continue the massage for a further five minutes in that position, making a total of ten minutes for massage.

6. If one is unable to retain the water for so long, one should empty out a part of the contents of the intestines and then take up the successive positions of the body and do the massages. 7. One should take only fruit and vegetable juices during the day on which the enema is taken. It is advisable, however not indispensable, that one take a sunbath and water bath immediately afterwards. 8. If the color of the liquid passed out is too dark, the enema should be repeated a second time.

THE IMPORTANCE OF ENEMAS

The enemas are necessary at the beginning of the treatment in order to get rid of all kinds of fermentations, harmful bacteria and by-products of the intestines and to prevent the new, wholesome foods from being fermented by the previous accumulations. During the cure, enemas are helpful from time to time to get rid of all the eliminated waste matter deposited in the intestines.

NATURAL DIETETICS AND THE FOODS OF LIFE

YOUR PRACTICAL GUIDE

At the beginning, this diet provokes unpleasant symptoms due to the general elimination of all accumulations of waste matter and toxins in the organism. It is usual for the old symptoms of past illnesses to be renewed. However, the individual who does not let himself be alarmed by them, but understanding the natural laws and trusting in the healing powers of the natural forces, perseveres in following the diet, will be abundantly rewarded by the cleansing and regeneration of his organism. So do not forget: every disease is curable, but not every person—only those who have strength of will and perseverance. Do not let yourself be influenced by passing symptoms of the cleansing process or by the opinion of ignorant people around you.

When you look at these proposals you may feel that it is is a complete change from all your usual means and customs, and therefore that it is too much for you. You may feel that you are going to lose too much pleasure for it to be worthwhile.

We can assure you that you will enjoy this different regime just as much and more, than you enjoy your present foods, and that your enjoyment will be increased by the knowledge that there will be no disagreeable after effects. Remember that all tastes are acquired. Once you lose the taste for meat, for artificial sweets, and harmful stimulants, it will never return to you. The more simple, natural foods you eat the more you will enjoy them.

If you feel that you can't begin with a complete break with your present habits, don't try. Try a little at a time. But never go back, keep on improving. For example, if you eat a heavy breakfast, begin to reduce it, cut down the variety of foods and the quantities, till in the course of two or three weeks you are down to a fruit drink. Then you cut that as well, and all the time you will gain benefit. If you are eating or drinking anything that is harmful, begin gradually to lessen the amount of it until at last you cut it out altogether. Remember that while you are doing this you are not only approaching the best diet, but also strengthening your will and self command and thereby building up confidence, a very important factor in dealing with illness and restoring health.

"Let food be your medicine. Let medicine be your food."
—*Hippocrates,* the Father of Medicine.

In the organic gardens at the Essene School of Life, around 1946.

An Essene picnic near Lake Elsinore, California, around 1936, close to the time of the writing of this book. Professor is in the center. At far right is Purcell Weaver, Professor's lifelong friend and translator of *The Essene Gospel of Peace* from French into English.

FOOD AS THE CHIEF INSTRUMENT OF HEALTH

MODERN DIETETICS AT A GLANCE

The Optimal Dietetic Utilization of the Proteins, Fats, Carbohydrates, Calories, Vitamins, Mineral Matter and Other Physico-chemical Properties Accumulated in 200 Health-giving Foods of the Vegetable Kingdom: The Pillars of Longevity

Category I

RAW JUICY FRUITS

STRONGLY ELIMINATIVE FOODS—40% OF THE DAILY RATION

Group A

1. Apple
2. Pear
3. Peach
4. Apricot
5. Plum
6. Prune, fresh
7. Nectarine

Group B

8. Strawberry
9. Raspberry, Black
10. Raspberry, Red
11. Huckleberry
12. Currant, Red
13. Currant, White
14. Currant, Black
15. Blackberry
16. Gooseberry
17. Cranberry
18. Mulberry
19. Medlar
20. Mirabelle
21. Tangerine
22. Grape
23. Cherry
24. Pomegranate
25. Black Fig, fresh
26. Watermelon
27. Canteloupe

Group C

28. Lemon
29. Lime
30. Orange
31. Grapefruit
32. Mango
33. Papaya
34. Persimmon
35. Pineapple
36. Custard Apple (Cherimoya)
37. Cactus Fruit
38. Cactus Leaves
39. Guava
40. Guava, Strawberry
41. Jujube
42. Durian
43. Sapodilla
44. Sapote, White
45. Star Apple
46. Litchi, Fresh
47. Litchi, Dried
48. Tamarind

Category II

RAW VEGETABLES

ELIMINATIVE FOODS—30% OF THE DAILY RATION

Group A

49. Tomato
50. Cucumber

Group B

51. Turnip
52. Turnip Leaf
53. Radish (large)
54. Radish (small)
55. Parsnip
56. Beetroot
57. Sugar Beet
58. Sugar Beet Leaf
59. Beetroot, Red
60. Kohlrabi
61. Carrot
62. Oxalis (Sorrel)
63. Artichoke
64. Jerusalem Artichoke

Group C

65. Lettuce, Romaine
66. Lettuce, Boston
67. Lettuce, Red Leaf
68. Brussels Sprouts
69. Cabbage
70. Cabbage, Savoy
71. Cabbage, Red
72. Cauliflower
73. Okra
74. Dandelion
75. Leek Bulb
76. Leek Leaf
77. Asparagus
78. Swiss Chard

Group D

79. Celery
80. Celery Root
81. Spinach
82. Watercress
83. Chickory Root
84. Green Pepper
85. Salsify
86. Salsify, Black
87. Kale
88. Rutabaga

CONDIMENTS WITH OIL AND LEMON JUICE FOR SALADS

1/5 OF THE DAILY RATION

Group E

89. Chive
90. Onion
91. Garlic
92. Dill
93. Parsley
94. Horseradish

Group F

95. Cow's Milk, fresh, raw
96. Goat's Milk, fresh, raw
97. Buttermilk
98. Yogurt
99. Skim Milk
100. Whey

Category III

WHOLE GRAINS AND STARCHES

RECONSTRUCTIVE FOODS—20% OF THE DAILY RATION

Group A

101. Banana
102. Potato
103. Sweet Potato
104. Breadfruit

Group B

105. Wheat, Whole
106. Wheat, Germ of
107. Wheat, Bran
108. Wheat Bread, Whole
109. Pumpernickel Bread
110. Swedish Rye Crisp

Germinated Wheat Grain

111. Whole
112. Crushed
113. Sun-Baked
114. Oven-Baked
115. Boiled

Group C

116. Rye, Whole
117. Corn, Whole Yellow (Maize)
118. Corn, Green (Maize)
119. Corn, Popped
120. Rice, Whole
121. Rice, Bran
122. Millet
123. Oats, Whole
124. Oats, Rolled
125. Oatmeal
126. Buckwheat
127. Barley, Whole

Group D

128. Sorghum
129. Sago
130. Tapioca
131. Taro
132. Yam
133. Cassava, Sweet

Category IV

DRIED, SWEET AND OILY FRUITS AND ANIMAL PRODUCTS

STRONGLY RECONSTRUCTIVE FOODS—10% OF THE DAILY RATION

Group A

134. Raisin
135. Prune
136. Fig, Dried (Smyrna)
137. Fig, Dried (Black)
138. Date
139. Apple, Dried
140. Apricot, Dried
141. Pear, Dried
142. Peach, Dried
143. Currant, Dried

Group B

144. Honey
145. Sugar Cane, raw
146. Pure Maple Syrup
147. Blackstrap Molasses

Group C

148. Almond
149. Walnut, English
150. Walnut, Black
151. Brazil Nut
152, Pecan
153. Butternut
154. Paradise Nut
155. Candlenut
156. Chestnut, Dried
157. Water Chestnut
158. Cocoanut
159. Cocoanut, Dried
160. Cocoanut Milk
161. Pignolia, Italian
162. Pistachio
163. Hickory Nut
164. Beech Nut
165. Piñon
166. Filbert
167. Peanut
168. Peanut Butter
169. Almond Butter
170. Carob
171. Cocoa Bean
172. Avocado
173. Olive, Dried
174. Sunflower Seed
175. Sesame Seed

Group D

176. Cottage Cheese
177. Cream, raw
178. Butter, raw
179. Egg, Whole
180. Egg, White
181. Egg, Yolk

Category V
COOKED BEANS AND VEGETABLES
CONSTRUCTIVE FOODS–SUPPLEMENTS TO THE 4 BASIC CATEGORIES

Group A (35%)

All Vegetables of Category II in cooked form, and the following:

182. Pumpkin
183. Lima Bean, Green
184. Kidney Bean, Green
185. String Bean, Fresh
186. Peas, Green

Group B (15%)

187. Eggplant
188. Soya Bean
189. Lentil
190. Mushroom
191. Champignon
192. Truffle

Group C (10%)

193. Bean, Dried
194. Bean, Blackeye
195. Horse Bean
196. Lima Bean, Dried
197. Kidney Bean, Dried
198. Peas, Dried
199. Cow Peas, Dried
200. Chick Peas (Garbanzas)

THE BASIC FOOD-CATEGORIES OF COSMOTHERAPY

Category I
FRESH JUICY FRUITS

This category is rich in vitalizing properties. It should form about one half (40%) of the total meal. All these juicy fruits may be taken in their natural form or their juice squeezed out, as long as the juice is taken fresh, immediately after preparing it. All the items given in this category are very efficient eliminators. They should be eaten *fresh.* This rule applies equally to all foods in all four categories without exception. These foods must constitute the whole of the first meals after a fast. This category constitutes an ideal in diet towards which we must strive; it is an ideal that can be realized in warm and temperate climates where such products, combined with a variety of nuts, are found. The foods in this category are the best conservers of perennial youth. They are the ideal food for the intellectual and sedentary worker, and ideal in general for all during the warm seasons and in warm climates.

Category II
FRESH VEGETABLES

They should form 30% of the total meal. They are edible either each item separately or in salad form—finely chopped or grated and mixed with lemon juice, oil and one of the condiments on the list. Milk is a vegetable transformed into milk in the organs of a cow, goat or sheep at a natural body temperature. As long as the milk is from a healthy animal who has never been treated by any drugs or injections, and who has pastured on unsprayed grass and leaves grown on rich soil untreated by commercial fertilizers, then it is alive as vegetables are, but has less power of elimination and more nutrition. As long as strict rules of hygiene are followed, milk and milk products should be raw, and fermented milk products, such as buttermilk, cheese and yogurt, should preferably be made at home.

Category III

WHOLE GRAINS AND STARCH FOODS

Whole grains are not eliminating, but building foods. They are chiefly muscle builders while their minerals contribute to the formation of the bones and their vitamins represent the best source of the Vitamin B complex. They are excellent foods particularly for manual laborers and sportsmen and for all in general in cold seasons and climates. Through muscular activity they are transformed into muscle fibre; through inactivity they become fatty tissue; taken in larger quantities than necessary they scarify the organism. They harmonize very well with Categories I and II, with which they form valuable combinations. They should form 20% of the total meal, except in the case of hard manual laborers, in which case they may be doubled because in the organism of this type of person every process of metabolism is stronger and more rapid than it is in persons of ordinary activity. The most valuable grain is wheat which contains in itself all the elements needed by the human organism. After wheat come all the other grains. To shorten the cooking time, soak whole grains overnight, rinse well and use the minimum amount of water necessary. Sprouting whole grains* adds immeasurably to the nutritional value and eliminates the need for cooking. The bread described in *The Essene Gospel of Peace* was baked in the sun and the only ingredient was sprouted wheat. Whether eaten as cereal, or bread, or in the form of sprouts, all grains should be eaten in their whole, integral form, with no part of the grain removed or processed in any way.

Category IV

DRIED FRUITS, NUTS, MILK PRODUCTS, EGGS

This category contains the most concentrated, heat producing and exceptionally nourishing foods. In larger amounts than required they hinder the process of elimination

***The Essene Way–Biogenic Living* and *The Chemistry of Youth,* among other books by Dr. Szekely, contain detailed sprouting techniques. Recipes for "Essene Bread", also called "Zarathustra Bread," appear in *Biogenic Reducing* and *The Book of Living Foods.*

and create fatty tissue in the organism. For this reason they should form not more than 10% of the total meal. They are excellent for cold seasons and cold climates and particularly for hard manual laborers in whose organism they are transformed into muscle fibre. Sedentary and intellectual workers need little of them and less during warm seasons and in warm climates than in cold. They harmonize very well with the other three categories. Honey is a certain constituent of flowers and fruits, transformed into natural sugar in the internal organs of the bee at a natural body temperature. Honey is as much alive, therefore, as the flowers from which it is derived, only far more concentrated. Its effect on the human organism is identical with that of the dried sweet fruits. Egg is a cereal transformed in the internal organs of the hen at a natural body temperature. It is, therefore, just as much alive as the grain from which it is derived, but much more concentrated. Milk—use the raw, unpasteurized product. Sour Milk—let the fresh milk stand two or more days until it clabbers, or use home-made buttermilk or yogurt. Cottage Cheese—put the clabbered milk into a cheesecloth to drip for one or one and a half days. Raw cream—syphon it out from the top of your (raw) milk bottle. Butter—beat the raw cream for a few minutes and it will turn into butter.

Category V
COOKED VEGETABLES AND BEANS

The foods listed here cause no elimination, but contribute building elements, proteins, calcium, phosphorus, iron and Vitamin B complex. We recommend their use as permanent supplments to the previous four categories. They are more advantageous to manual laborers than they are to sedentary and intellectual workers. They should be soaked in water overnight before cooking (or sprouting), and revitalized by mixing them with oil, lemon juice, parsley, onion and other condiments. The soya bean is a perfect alkaline protein, equivalent to nuts.

Professor with a group of students on an expedition in the Swiss Alps, circa 1930.

THE TWELVE GOLDEN RULES OF A PERFECT DIET

Eat every day some raw fruits, vegetables, and sprouts. Raw foods contain essential vitamins and enzymes.

Eat only natural foods; see that your fruits and vegetables have not been contaminated with pesticides or cultivated with commercial fertilizers. To ensure this, it is a good plan to unite in groups which collectively obtain natural foods direct from trustworthy producers.

Eat simply: two, three, four, or at the most five kinds of food at one meal and mistrust too complicated combinations. Our gastric juices easily digest two or three different substances, but digest with difficulty or incompletely too complicated combinations.

Avoid the eating of meat, fish, fowl and crustacea. Flesh foods cause intestinal putrefaction, unfavorable intestinal flora and contribute to general toxemia. Eat natural cheese, fresh eggs, raw milk, nuts and seeds instead.

Eat with moderation, and in small amounts, according to individual needs, mode of life, season and climate. You may eat less, but not more. Always listen to the voice of your organism. Everything superfluous in your organism accumulates waste matter and causes overweight and fermentations, hindering the activity of the heart and circulation. All the long-livers who exceeded one hundred and twenty years ate with moderation—a small amount which when perfectly masticated is amply sufficient for the needs of the body. More people die of overeating than of starvation. The more rapidly you eat, the more you desire, and the more you consume. The greater the superiority of a food, the less of it is required; therefore, select your foods according to quality rather than to quantity.

Eat only three times a day at most, according to your own needs and mode of life. The system of not more than three meals a day lengthens life.

The fruits and vegetables should by preference also be the product of the climate and region where you live. Local produce is more suited to the organism.

Food should be masticated very thoroughly. It is important to put very little in the mouth at a time, to chew every mouthful thoroughly until it is reduced to a liquid before swallowing, not to take a fresh mouthful until the previous one has been swallowed, and to mix food very thoroughly with the saliva in the mouth. These rules are important because foods not sufficiently masticated or mixed with the saliva remain partly undigested and therefore unavailable as nourishment for the organism.

Never eat without appetite; never eat when in bad humor; never eat when tired. This is another vital rule because if your nerves are not in equilibrium owing to fatiguing or discordant thoughts, feelings or actions, they cause disturbances in your glandular secretions during the process of digestion. Therefore, if you are tired, perturbed or emotionally upset, refresh yourself before eating with a rest, walk, breathing exercises, bath, reading, etc., until you regain the state of calm most favorable to healthful eating.

Fast with water or fresh fruit juices, one day a week. Fasting is an efficient remedy for diseases, an eliminator of waste matter and a great force for the regeneration of the organism. When the organism is occupied with the processes of digestion its force is limited in the performance of other activities and is therefore unable to eliminate in full strength. During a fast the vitality of the organism is liberated and it then brings about the elimination of waste substances in the body. Fasting is also capable of compensating for occasional over-indulgence in food. So it is especially recommended after such abuses.

Even for healthy persons fasting is very regenerative. During a fast you must keep the body at a good normal temperature; you must be moderately active; you must exercise in sunshine, air, and water; you must breathe pure air rhythmically, and refresh yourself by washing of the skin.

Eat preferably those foods which are riches in minerals, vitamins, enzymes and hormones. Mineral, vitamin, enzyme and hormone deficiencies are the principal cause of the deterioration of the basic biological functions of the organism.

HARMONY WITH NATURE'S CYCLES
DIET ACCORDING TO SEASONS

Foods	*Months*
Milk and Milk Products (of Categories II and IV)	March, April, May
Juicy Fruits (of Category I)	June, July, August September, October, November
Cereals, Dried Fruits and Nuts (of Categories III and IV)	December, January, February
Salads and Vegetables (of Category II, same as Category I)	

NOTES

The foods indicated above are not the exclusive foods for the period, but simply the foods favorable for each season.

Naturally, the foods should also be modified in accordance with the demands of the individual.

THE TEN COMMANDMENTS OF A HEALTHY DIET

Foods Entirely Prohibited	*and Substitutes Recommended*
All Flesh Foods	Eggs, Natural Cheese, Beans, Nuts, Seeds, Fermented Milk Products
Tobacco	Fresh Air
Alcohol	Fresh Natural Fruit or Vegetable Juices
Coffee	Fig, Cereal, Soya Bean or Dandelion Coffee with Warm Milk and Honey
Tea	Natural Herb Teas (e.g. Mint, Elderberry, Camomile, Lime Blossom), with Honey and Lemon Juice.

White Sugar	Honey, Pure Maple Syrup, Dried Fruits
Refined Flour Products	Sprouted Whole Wheat Bread, Other Whole Grains, Preferably Sprouted.
Chocolate, Candies, etc.	Dates, Figs, Raisins, and Other Dried Fruits
Salt	Onion, Garlic, Celery, Dill, Other Herbs and Dried Vegetables
Vinegar	Lemon and Grapefruit Juice

THE TEN REASONS FOR THE TEN COMMANDMENTS

The Reasons Why These Foods are Prohibited:

All flesh food (meat, fish, fowl and crustacea (1) introduce into the organism uric acid and other harmful acids; (2) cause intestinal fermentation and putrefaction by their decomposition in the organism; (3) introduce into the organism many parasites from dead animals; (4) introduce into the organism diseases from dead animals, as well as all kinds of antibiotics, chemicals and hormones injected into the animals in attempts to prevent disease; (5) are extremely hard to digest, taking a great toll on the organism in both time and energy; (6) have an exciting influence on the nervous system; (7) furnish a favorable medium for the culture of the bacteria of disease; (8) lessen the resistance of the body to disease.

Nicotine is a poison. It (1) decreases the capacity of oxygen-absorption of the pulmonary cells; (2) prevents the organism and particularly the blood from obtaining the maximal amount of oxygen; (3) lowers the quality of the blood; (4) causes nervous pains; (5) progressively and increasingly attacks the heart; (6) hinders mental work; (7) injures the eyes; (8) causes lung cancer; (9) in a pregnant woman, gravely threatens the life and health of the fetus.

Alcohol (1) ferments in the organism and causes fermentation of the food with which it mixes in the stomach; (2) progressively attacks the nerves; (3) contributes to high blood pressure; (4) paralyzes mental activity; (5) degenerates the character of the individual; (6) may severely deteriorate certain parts of the body, especially the liver and cells of the brain; (7) in a pregnant woman, gravely threatens the life and health of the fetus.

Coffee and Tea attack the nerves and the heart through their caffeine and theine content. In a pregnant woman, caffeine and theine may threaten the life and health of the fetus.

White Sugar contributes no mineral or vitamin value to the organism and consequently causes mineral deficiencies. It contributes to the deterioration of the teeth and may cause hyperactivity and other behavioral disturbances in children.

All refined flour products are superfluous, inferior starch in the organism, without mineral and vitamin value. Because of the missing fibre content in refined flour products, they contribute not only to constipation and other intestinal problems, but also to cancer of the colon.

Chocolate is (1) constipative; (2) deposits in the organism refined, demineralized and devitaminized substances of only caloric value, as well as deteriorating the teeth.

Salt (1) through progressive accumulation contributes to arteriosclerosis; (2) irritates the mucous membranes; (3) contributes to high blood pressure; (4) provokes unnecessary thirst and water retention in the tissues.

Vinegar (1) causes inferior fermentations in the organism; (2) irritates the mucous membrane.

Note: When this book was first written, the billions of pounds of chemicals added to commercial food products every year had not yet appeared on the scene. Now, in addition to the above-mentioned harmful foods, we must contend with artificial colors, synthetic additives, preservatives, humectants, anti-oxidants, emulsifiers, artificial flavors, etc. etc. etc. All these additives are extremely dangerous and may even be carcinogenic; there is no such thing as a "safe" or "harmless" quantity of these chemical substances.

DIETETIC APPENDIX NO. 1

THE ESSENTIALS OF AN ADEQUATE DIET AT A GLANCE

Protein	Builds and repairs all body tissues.
Calories	Furnish heat and energy.
Minerals	Build and maintain the body and regulate its basic biological functions.
Calcium	Builds and maintains strong bones and teeth and normal body tissues.
Phosphorus	Builds and maintains strong bones and teeth and normal body tissues.
Iron	Builds blood and protects against nutritional anemia.
Copper	Aids in utilization of iron by the body.
Other Minerals	Aid in various body functions.
Vitamins	Promote growth and maintain health.
Vitamin A	Aids in normal functioning of the eyes and increases body resistance.
Vitamin B1 (Thiamin)	Stimulates appetite, prevents beri-beri, aids in utilization of carbohydrates, and promotes the health of the body cells.
Vitamin B2 (Riboflavin)	Is necessary for cell oxidation and protects against nervous disorders and liver disturbances.
Nicotinic Acid	Prevents and cures pellagra.
Vitamin C	Prevents and cures scurvy and aids in the maintenance of healthy body tissues.
Vitamin D	Aids in utilization of calcium and phosphorus for the prevention and cure of rickets.
Vitamin E	Plays very important role in the functions of the sexual glands and in the activity of the reproductive systems.
Bulk	Aids in the elimination of body waste.
Juices and Water	Help to flush accumulated waste-matter out of the body.

DIETETIC APPENDIX NO. 2

OUR SIMPLEST AND QUICKEST DIET PLAN FOR THOSE PREPARING MEALS

There are eight basic food groups, each of which supplies vital food elements necessary for buoyant health and strength—vitamins, minerals. enzymes and high grade proteins. It is essential that your daily diet contain at least one food from each and every one of these eight basic groups.

I

Milk and milk products. . . at least a pint a day for everyone—more for children—or its equivalent in cheese or other milk and fermented milk products.

II

Oranges, tomatoes, grapefruit, sprouted seeds and grains of all kinds, raw cabbage or salad greens. . . if possible, a cupful of orange or grapefruit juice; or a large helping of tomatoes or raw cabbage or salad greens.

III

Green or yellow vegetables. . . at least one big helping: some raw, some cooked.

IV

Bread, cereals. . . choose whole grain products or whole grain bread.

V

Dried beans, peas and nuts.

VI

Eggs. . . at least three or four a week, cooked any way you choose—or as a part of other dishes. Make sure the eggs are from hens who are allowed to scratch freely and live in a natural way, outdoors.

VII

A small amount of unsaturated fat, preferably in the form of avocado or raw sunflower seeds. Unsaturated vegetable oils can become unstable very quickly, and then they are harmful to the organism.

VIII

This final group includes potatoes or other fresh tubers and other vegetables or fruits. . . to round out your diet with plenty of variety.

On top of this foundation of the eight basic food groups of health and strength, you may eat other foods you also like, as long as they are not refined, processed or chemicalized, and are not on the list of prohibited foods.

THE REWARD OF YOUR PERSEVERANCE

THE REGENERATIVE PROCESSES IN THE BASIC BIOLOGICAL FUNCTIONS OF THE ORGANISM UNDER THE COSMOTHERAPEUTIC WAY OF LIFE ADVOCATED IN THIS BOOK

I. *In the general condition of the organism:*

–Eliminates from the organism the toxins and superfluous matter accumulated through following an unhealthy diet;

–Rejuvenates the organism beginning to lose its vital force through age;

–Assures deep, quiet and refreshing sleep in case of sleeplessness;

–Strengthens the organism in case of general weakness caused by disease or exhaustion of the physical forces;

–Refreshes the body in case of fatigue due to exaggerated activity;

–Eliminates the excess of liquid from the organism in progressively increased measure;

–Eliminates the excess of fat from the organism through dissolution and reduction;

–Reduces in case of excess weight, in a healthy manner, without weakening the body and without disagreeable consequences;

–In case of weakness following on defective diet strengthens the organism by creating new muscular cells and builds up again the natural weight of the body.

–Eliminates all excess of acidity from the organism by accelerating the internal and external secretions;

–Eliminates gases from the system by rhythmical movement, thermic and biochemical effects;

–Relieves acute and chronic pains by dissolving and eliminating accumulated toxic matter;

–Diminishes and gets rid of fever by reduction of excess heat;

–Gets rid of chills and fever by restoring the normal internal heat of the organism;

–In case of local insensibility progressively increases the local metabolism;

–Eliminates and stops fermentation and putrefaction in the intestines through internal cleansing;

–Softens scarification, sclerosis and other hardenings in the organism;

–Improves troubles of the processes of metabolism, accelerates and intensifies metabolic activities in the organism;

–Refreshes and strengthens the nervous system by eliminating the physiological causes of nervous tension and irritation;

–Relieves, lessens and progressively puts an end to nervous troubles, tensions and blunting of the nerves by psychological diet in thoughts and feelings;

–Ends passivity, pessimism and lack of will, giving a new, healthy, robust and harmonious life-rhythm to the organism;

–Stops exaggerated appetite by introducing natural energies into the organism and making alimentary excess unnecessary;

–Gets rid of abnormal or exaggerated sexuality by eliminating excess of energy and internal sensorial excitants;

–Ends sexual debility (impotence or frigidity) by reactivating the glands of the organism;

–Renews the blood, changing it into a vitalizing liquid which improves the whole organism by its circulation;

–Cleanses, improves and makes more elastic the cells and tissues of the organism;

–Stops the processes of deterioration of the basic biological functions of the organism;

–Improves the basic biological functions of the organism.

II. *In the epidermis:*

–It makes the skin finer, stronger and more elastic, and a healthy skin with adequate respiration and perspiration facilitates the functioning of the lungs and kidneys.

III. *In the lungs and respiratory system:*

–It increases the capacity of oxygen absorption of the lungs and oxygen transport of the circulation of the blood, and by this accelerates all the metabolic processes of the organism.

IV. *In the heart and circulation of the blood:*

–Balances and improves the chemical composition of the blood, deficient, as a result of inaccurate diet, and makes the new, purified, enriched blood carry out a permanent activity of dissolution and elimination in the organism; it makes the pulse and movement of the heart easier and more regular, improves blood count, increases low and decreases high blood pressure.

V. *In the digestive and excretive system:*

–It purifies and revitalizes the cells and appropriate glands of the stomach, liver, pancreas and kidneys; it improves the dissolution, assimilation and distribution of food substances in the organism; banishes excess or lack of stomachic acidity, stomachic and intestinal fermentations, intestinal gas and intestinal sluggishness.

VI. *In the muscular system:*

–It strengthens and makes more elastic the muscular cells, fibres and membranes; it builds new muscular cells; normalizes the contraction and dilation of the muscles and the whole circulation.

VII. *In the sexual and reproductive organs:*

–It gets rid of local inflammations and swellings of the male and female sexual organs, making the function of the organs balanced and periodically rhythmical; it frees menstruation from disagreeable, accompanying symptoms, making it regular and definitely dividing days when conception is possible from those that are sterile, and by these means makes birth control easy, certain and natural, without the use of artificial, harmful contraceptive appliances or pills; it assures the most favorable development of the processes of conception and gestation; it reduces the pains of childbirth to the minimum and eliminates the usual complications; it increases the vital strength of the embryo and assures a favorable heredity to the new-born; it gives revitalizing and constructive force to the mother's milk and gives the infant a strong constitution, strengthening his capacity of resistance to childhood diseases.

VIII. *In the bony system:*

–It contributes to the calcium-phosphorus reserve of the bony matter of the skull, skeleton, back-bone and extremities and favorably influences local deformities.

IX. *In the nervous system:*

–It increases sensibility of nervous reception and intensity of nervous reaction; it relieves nervous pains and tensions with their derived symptoms and reflexes; it improves the nerve cells and the specific capacities of the nerve center and nerve endings of sight, hearing and feeling.

X. *In the internal and external system of secretory glands:*

–The reactivation and bringing to a state of mutual equilibrium and harmony of the endocrine and exocrine secretory glands, improving the basic biological functions of the whole system.

Professor Dr. Edmond Bordeaux Szekely in 1928.

THE GOLDEN RULES OF LIFE AND HAPPINESS

THE PHILOSOPHY OF NATURAL LIFE IN THE TEACHINGS OF THE GREAT MASTERS OF HUMANITY

Simple, natural nutrition in moderation is a saving of time, money and strength. It increases bodily and psychic energy and strengthens resistance to disease.

—*Cornaro*

Air is the most important nutriment.

—*Hippocrates*

Sunshine is the best remedy.

—*Plinius*

Water cures every illness.

—*Epicurus*

Vitality and beauty are gifts of nature for those who live according to its laws.

—*Leonardo da Vinci*

Health and Happiness are the Measure of all things.

—*Epicurus*

Happiness is not easily found. It is difficult to find it in ourselves and impossible to find elsewhere.

—*Chamfort*

Everyone must begin with himself in order to create his own happiness. Undoubtedly in this manner the happiness of all mankind will finally be realized.

—*Goethe*

Evil exists not: only the past. The past is past, the present is a moment, the future is all!

—*Zarathustra*

THE TWELVE GOLDEN RULES OF HEALTH AND HAPPINESS

Air, Water, Sun, Exercise, Respiration, Altruism, Culture, Harmony

1. Spend as much time as possible in the fresh air.
2. Always take a cold or lukewarm shower in the morning when you wake and in the evening before going to bed.
3. Devote one half-day a week to alternate sunbaths and waterbaths (swimming where possible).
4. Go for a thirty to fifty-minute walk in the fresh air after lunch and dinner every day, breathing deeply and rhythmically.
5. Do some simple physical culture exercises for 10-15 minutes every morning on rising.
6. Always sleep in a cool, airy and quiet place.
7. Follow the middle path of moderation and naturalness in all things, particularly in food, rest, work, recreation and sleep.
8. Beware of psychological toxins (inharmonious thoughts and feelings) even more than of chemical poisons. Be an optimist and have confidence in yourself, in human progress and in the healing forces of nature.
9. Be a sincere altruist, love and help others. Appreciate people's good qualities and forgive their bad ones.
10. Enjoy the beauties of nature and the masterpieces of literature, the arts and philosophy.
11. Study and observe the natural and cosmic laws and live in complete harmony with the natural and cosmic forces.
12. Do everything possible to spread the light, in order that the darkness might dissipate. Be an active point in the universe.

GENERAL HYGIENE, THE PRESERVER OF HEALTH

A rational dietary is not enough to keep in health. We must observe the laws of hygiene. We must have a passion for cleanliness, we must use plenty of water and wash all parts of the body thoroughly in order to ensure that the skin shall perspire freely and be kept supple and healthy. Cold water closes the pores, warm water opens them. Cold baths give strength and vigor, lukewarm baths soothe the nerves and dispel fatigue; excessively hot baths are enervating and weakening.

Leave the windows of your room as wide open as you can so as to let in as much light and air as possible. Rooms without fresh air and cross-ventilation are breeding-places of disease. In rooms with stale air, colds, bronchitis, pneumonia, sore throats and influenza rapidly develop. Let the song of the wind come in, the wind which sweeps out your house and brings the daily bread of your lungs upon its wings.

REGULAR AND COMPLETE ELIMINATION, THE PREVENTATIVE OF DISEASE

Ninety percent of our diseases originate in the intestinal system, including intestinal toxemia, putrefaction and fermentation.

Emptying regularly the intestines prevents a host of serious illnesses: nephritis, dyspepsia, enteritis and appendicitis follow upon constipation. So we should make a habit of allowing our bowels to function regularly. The bowels, like the stomach, like regularity and readily accustom themselves to work at a particular time of day. Help them to be punctual, preferably in the morning, by going to the bathroom whether you feel like it or not and they will soon fall into their routine and obediently conform, provided you follow the right diet and do daily walking or exercises. There are natural ways to overcome constipation, through roughage, juices, and exercise.

RIGHT CLOTHING FOR COMFORT

Our clothes should be comfortable, non-restrictive, and

adapted to the climate. In cold weather, the thickest are not the best. Two light garments on top of one another retain heat better than a single heavy one. Air is a bad conductor of heat, and so the heat of your body is retained better if there is a layer of air between the various things you wear. Try to choose natural fabrics, such as cotton and wool, and avoid synthetics. In warm weather, do not let the dictates of fashion overcome common sense.

Pay attention to your footwear. There is no more excuse for anyone not to wear truly comfortable, supportive shoes, because they are now so readily available, thanks to the fact that walking and running have become so popular with everyone, not just professional athletes. High heels, for whatever reason, are to be avoided. Sandals in summer should be supportive to the foot, and if large grassy areas can be found, there is nothing healthier than to walk barefoot on grass, especially wet grass in warm weather.

SOBRIETY, THE SICK MAN'S SAVIOUR

In the period of the Renaissance, the Venetian nobleman Luigi Cornaro was a model of abstemiousness. At the age of forty he was near death and suffered a great deal. His doctors—and they were many—were unable to cure him. In the end he took up the method of a Milanese doctor who was a disciple of the school of Salerno which was based on natural laws. From that moment Cornaro took only twelve ounces of food and fourteen ounces of liquid a day. His health soon returned. But Cornaro continued to follow the diet because he thought that if it could give him back his health it would certainly preserve it, and he lived in this way to the age of 78. So great was the improvement in his health that he outdid much younger people at riding and walking. When he was 78 his friends induced him to increase his ration by two ounces a day. His organism, which was perfectly accustomed to his previous exact ration, could not stand this addition. He fell ill again, so ill that he was forced to reduce his original allowance by three and a half ounces a day. He recovered his health. With a daily ration of about

8 ounces of food and 11 ounces of liquid he lived to be 104. He wrote four books on sober living, the most interesting of them being written when he was 96. It is a pity that he did not know the laws of natural diet, as with his remarkable perseverance he would certainly have been able to prolong his life by another twenty years.

COMPLETE MASTICATION, THE HEAVY EATER'S SOLUTION

The question of thorough mastication is of very great importance. A large number of heavy eaters have managed to reduce the amount of their food in this way. The digestion of foods has already begun in the mouth. The function of the saliva is to set going the processes of dissolution and assimilation. The gastric acid is only able to dissolve into its constituent elements food which has been well masticated and mixed with the saliva. If we masticate completely, our appetite is satisfied after eating a very small quantity. We should chew each mouthful thoroughly, depending on the nature of the food. It is the tongue which measures the satisfaction of our appetite. The tastes of foods are absorbed from the tongue by our nervous system.

WALKING, THE ROAD TO FITNESS

Make the most of your leisure moments and devote at least one hour of your day to walking. It is the best gymnastic exercise. In popular terminology, walking "oils the joints." It prevents stiffness and atrophy, keeps the body in good condition and gives us suppleness and harmony in our movements. What could be more awkward than a large, ungainly stomach! Besides, excessively fat, breathless, apoplectic, bulky people are liable to a host of ailments. A daily walk will keep you thin and will have a great influence on breathing, circulation, nutrition and elimination. It will keep your body well proportioned, while your muscles, toned up and free from excessive fat, will preserve their strength and elasticity even in old age. Besides being good for you physically, it is none the less so morally. Shakespeare is neither the first nor the only one to have discovered the connection between walking and a calm conscience. It makes the mind clear and ordered, dispels resentment, inspires good resolutions. Writers and artists do their best work while walking.

You should walk quickly, taking fairly large steps, with the ball rather than the heel of the foot resting on the ground. A conscious effort should be made to swing the arms.

THE ART OF BREATHING

Learn to breathe. Hardly anyone understands this art, which all should know and practice. It is the air which renews our blood and brings life to all our organs. It is the air which help to give us balance and to keep our physical and psychological functions in good order.

The expansion of our lungs by a simple mechanical movement opens out the seventeen or eighteen hundred million air cells of which they are formed. If the cells were placed side by side they would cover an area of more than a hundred and forty square yards. According to the best physiologists, most people only use a third, a quarter, or even a fifth part of the lung's total surface. In other words, if their air cells were laid out flat, they would hardly fill a space of forty-six, thirty-five, or twenty-eight yards, as the case may be. We can appreciate from this what a small amount of air we absorb in comparison with the quantity theoretically available.

We should use as many of the cells as possible. The way to do so is to fill the lungs to their utmost capacity by learning to breathe deeply. Our clothes should be loose enough to allow the thorax to function freely. Inhale for about seven seconds, mentally counting, one, two, three, four, five, six, seven, to measure the time; count three in the same way for a rest and then exhale for seven seconds, counting as before. Always breathe through the nose, with the mouth shut. The first exercise should last ten minutes. As you gain experience, you can progress to breathe more and more perfectly, cultivating a slow, regular, deep rhythm which will make the air reach the minutest folds of your lungs and charge your blood with life-giving oxygen, source of a feeling of fitness and good health.

The most important precondition to attaining the maximum benefit from rhythmic breathing is, obviously, the quality of the air. Breathing exercises should be done only in pure, unpolluted air. If you are lucky enough to be in the country where the air is filled with the fresh scent of fields and woods, then do the exercises while walking at a steady pace. You should also do them every morning in your room with open windows. Perseverance will be rewarded by an exceptional feeling of vitality and self-control.

EDUCATION, THE INSTRUMENT OF PROGRESS

Education is not the least of our requirements. All the foregoing constitute education of a kind which is essential. Bodily hygiene is the pre-requisite of intellectual progress. It will brighten the mind, relieve depression and arouse the desire to know more. This desire should be encouraged by systematic study and thoughtful reading. Such activity is more agreeable and fruitful if pursued together with those who are similarly interested. Discussion and study in groups or circles should be encouraged because it develops social outlook and responsibility.

RECREATION AND THE PLEASURES OF LIFE

Recreation requires a change of interest, mental and bodily. Whatever your daily occupation may be, find another one to occupy your leisure hours.

Remember that health requires the development of both physical and mental faculties, therefore organize your leisure accordingly. Choose an interest; a hobby or study for your leisure hours which is agreeable to you will do much good.

Except in cases of fatigue or exhaustion, or when loss of sleep has to be made good, a day in bed or some hours lying down are definitely of less value than hours in the fresh air, in exercise, gardening, or in an agreeable and healthy recreation.

RELAXATION, THE MODERN MAN'S SAVIOUR

If regular work, suitable, proportioned to the worker's strength, so that overstrain and excessive wear and tear are avoided, is undoubtedly a prime factor in health, it is equally clear that to keep the balance and maintain the human machine in a satisfactory condition the worker must respect the physiological law of rest.

Relaxation of the muscles is essential. Even the slightest contraction of them gives the cells work to do. Breathe rhythmically, lower your eyelids, rid yourself of every anxiety and think of something pleasant, a lovely country scene, a forest, or a lake at sunset. Let the minutes glide by like a flowing stream and let your thoughts wander in a doze betwixt waking and sleeping.

Do not resist these few moments of vegetative life and do not try to spoil their good effects. After ten or twenty minutes you will get up fresh, calm, braced and eager for movement and effort. The relaxation will have repaired the wear and tear of your vitality and will have oiled the wheels again.

Every kind of abuse is a mistake, including abuse of rest. Though proper use of rest refreshes, an excess of it is more than likely to lead to illness before long. Inaction and laziness impair health even quicker than over-work. It is better to work too hard than not at all.

HARMONIOUS SLEEPING, THE SOURCE OF LONGEVITY

From relaxation it is only a single step to sleep. The quality of one's sleep is very important to health, as is the quantity. Insufficient sleep predisposes us to illness. Nervous exhaustion and weakness are often due to lack of it.

Sound sleep is the best rest. The windows should be open and the covering light and warm. You will have better sleep if you avoid heavy and close night attire. The hours before midnight are the most valuable. Regularity in retiring and rising helps the body. It is essential to calm the mind before going to sleep—a bath or a few exercises will do this. Don't begin to think on disharmonious things when you lie down or you will provoke dreams and a disturbed night.

You should not go to bed until at least two hours after dinner. Your bedroom should have an agreeable temperature and be well ventilated. Adopt the good habit of breathing through the nose and you will avoid dryness of the throat and snoring. Forget all about the bustle of the daytime; sink into a state of complete passivity and sleep will come of its own accord.

In the morning do not wake up abruptly or jump out of bed in a hurry. Give conscious life time to reassert control of your organs; give the brain a chance to brush away the cobwebs of sleep. Stretch yourself, lie quiet for a few minutes and then get up. A sudden awakening, a hasty start of the day, overstimulate the circulation and are a shock to the nerves. These jars to the system are bad for the health and should be avoided.

HYGIENE IN THOUGHTS AND FEELINGS: THE SOURCE OF YOUTH

"Man does not die; he kills himself," say all great thinkers of humanity. Human life is something more than the sum of the mechanical forces and chemical combinations at work in our body. Above and beyond matter there is a superior force directing and influencing the former, an essential principle, a vital dynamism, variously named at different times and by different philosophers. This must be our guiding principle in the regulation of our physiological and psychological functions. To obtain health and youth we must will it. It is a truth for which there is more than sufficient evidence, that a sudden outburst or passion in a hitherto calm and balanced individual is enough to sow the seeds of disease. Grief or fear can have a considerable effect on the system. Hope and joy bring the dying back to life, they warm and rejuvenate the blood and re-invigorate the muscles. We see, then, that thought can affect the body. We have within us the fountain of youth which makes man perennially young and strong and enables him to reach, without infirmity or decline, the limit of life assigned to us by nature. We think it extraordinary to live to be ninety. We ought to live to be more than a hundred and twenty.

What must we do to keep our precious youth to an advanced age? We must not only live abstemiously and observe the laws of health, we must also practice mental hygiene. Keep a watch on your thoughts, brush aside all feeling of depression and discouragement, and close the door to doubt, morbid fancies and every kind of phobia on the question of age and illness which could disturb your peace of mind.

The discipline of the inner life, able by progressive effort to brush aside the passions which bring about man's downfall, this calm philosophy that teaches us to master our impulses, the violent upheavals of our soul, has been known throughout the ages. We find it among the Indian Yogis, Essene Sages and Greek Philosophers. Seneca gave good advice to the neurasthenics of Rome when he wrote to his friend Lucilius: "The troubles of life and pain are easy to bear when they are not exaggerated by dwelling on them. Thinking makes it so."

Any kind of psychological disturbance means lowered resistance. Though we do not die of grief or fear, they affect our system in such a way that it easily falls prey to the enemy at the gate. The depressing effects of physical conditions, of climate, and daily intercourse with unbalanced people must be combated by an unshakable optimism. We succeed in this by cultivating the will. Not an imperative will implying effort, but a reasoned, permanent will which has become an instinctive reflex act. We must master life and dispassionately control our actions. This is the way in which we can, without being involved in a sterile egotism, ease our great sorrows, suppress anger, eliminate from hatred all but the element of indignation that drives us to action, and know the joy of forgetting and forgiving. Here lies the secret of perpetual youth.

TRIUMPH OVER DEATH

This peace of mind and soul will be a sure shield against disease and old age.

Buffon has said that what destroys balance in a man is an attitude of mind which dwells on evil at the expense of good. If you are not afraid of death, death will retreat before you. Most people, when they reach a certain age, think that they have come to the end of their lives. Just a few sad years of the discomfort and decrepitude inseparable from old age, and that will be the end. Their minds are filled with gloomy forebodings, their horizons are narrowed. They lose their buoyancy and are haunted by the approach of death. They are victims of the worst of auto-suggestion and cannot react. In their distress they eat badly; nutrition is impaired, their sleep disturbed, and their poor nerves, frayed by all these harmful thoughts, upset the whole system. The least symptom is seized on by their disordered imagination. They brood on it until it develops into the disease which will ultimately kill them.

Apart from violent unforeseeable accidents, death is always "real suicide or rather murder committed by the soul upon the body through ignorance or folly." We must defend ourselves against evil thoughts and feelings which shorten life and we must learn to keep in good spirits, to treat death with contempt and to cultivate the calmness of spirit which breeds centenarians.

A picture of Essene health: one of the young students at the Essene School of Life in Baja California, Mexico, around 1946.

APPENDIX

THE MODERN ESSENE RENAISSANCE

Never before during any other age of universal history has humanity been faced with such a great need to return to the ancient traditions of the Essenes, to a natural, simple, healthy and creative life, than during the present world chaos.

The seven principal reasons that prove the great necessity for a modern Essene Renaissance are:

1. The Essenes were the greatest agriculturists of Antiquity. They produced on comparatively poor soil (around the Dead Sea in the Holy Land) a surprisingly large variety of the best quality fruits, vegetables and cereals in a completely natural way. They did not use any artificial fertilizers or sprays as they knew the secrets of the best proportions of water, sunshine and the ingredients of the earth and atmosphere. We are greatly in need of renewing their natural and rational agriculture because today everything is grown on a poor, thoroughly exploited and impoverished soil, filled with artificial chemical fertilizers and sprays, the result of which is that ninety percent of our vegetables and fruits are deficient in minerals which are vital to the body. This fact is one of the most important sources of contemporary diseases, for which the only remedy is to return to the natural, rational Essene agricultural system.

2. The Essenes of the Dead Sea and the Therapeutae of Lake Mareotis in Egypt were the greatest healers of Antiquity. They evolved a perfect coordination between all the material and psychic forces, resulting in a wonderful dynamic synthesis of Health, Healing and Harmony. No better remedy can be found for the prevention and elimination of today's cauldron of diseases which inflict suffering upon ninety percent of the human race.

3. The Essenes lived a simple, natural and creative life in their self-subsistent agricultural Brotherhoods, far from the chaos of the big cities, and from all the destructive currents of violence, egoism and ignorance. They satisfied all their needs with only four hours work, four days a week, reserving their free time for meditation, study and leisure. Their simplified, natural form of life even permitted them to distribute annually all their surplus products among the poor. This may be compared with our present world era of artificial city life, nervous tension, economic crises, wars and disharmony.

4. The Essenes created the greatest synthesis of religions, philosophies and sciences of human history on the basis of the organic unity of the Hebrew, Egyptian, Greek, Persian, Babylonian and Hindu cultures. This harmonious synthesis inspires us to unify and reconcile all the values of universal culture in an optimal and omnilateral contemporary synthesis.

5. Modern scientific research has completely verified the scientific foundation of the ancient traditions of the Essene Science of Life, thus establishing the common basis of science, ancient and modern.

6. The basic Essene principle of creative love, taught by the healers and teachers from the Brotherhoods, amongst whom were Elijah, John the Baptist, John the Beloved and the great Essene Master, Jesus, is the eternal foundation of all religions and ethics.

7. It is our deep and sincere conviction that the renaissance of the original purity, simplicity and harmony of the free, creative life of the Essenes is the optimal solution for the present world chaos, violence, ignorance, misery and suffering.

Only upon this universal and eternal basis becomes possible the omnilateral realization of the best forms of individual and social life. Only thus can we obey the principle of practical altruism: to dispel evil not by attacking it but by maintaining and strengthening the good. To overcome ignorance with education, egoism with altruism, violence with constructive social activity, misery by the removal of its causes, disease by the strengthening of the organism's vitality and chaos by the organization of order. To create a better future, not by the destruction of the past, but by the constructive and progressive amelioration of the future.

Original pencil sketch by Edmond Bordeaux Szekely, date unknown.

APPLICATION FOR ASSOCIATE MEMBERSHIP
INTERNATIONAL BIOGENIC SOCIETY

Please return to: *I.B.S. Internacional*
Box 205, Matsqui, B.C.
Canada V0X 1S0

*Name*____

*Address*____

*City*____________*State/Prov.*__________*Zip/Code*________

*Age*___*Profession*____________________________________

*Previous Experience*___________________________________

I am interested in:

_____*becoming an Associate Member of the I.B.S.*

_____*becoming a Teacher of Biogenic Living.*

Enclosed is my annual Associate Membership fee of $20.00. Please mail my membership card, your next issue of the Periodical Review, *The Essene Way*, and my copy of *The Essene Way—Biogenic Living*, my most important textbook and encyclopedia of ancient wisdom and modern practice. I understand I will receive a 20% discount on all publications as an Associate Member, but only if I order *directly* from I.B.S. Internacional.

Please make your check in U.S. currency out to
I.B.S. INTERNACIONAL.

RECOMMENDED BOOKS FOR STUDY

Many members who cannot attend the International Seminars because of distance or limited means are interested in a systematic program of home study. The following books are recommended for such a program, and provide an excellent foundation for study of *The Essene Way of Biogenic Living*, especially when coordinated with the methods outlined in *The Art of Study: the Sorbonne Method.* Also, it is recommended that these books be read and studied before attending the *International Seminars on the Essene Way and Biogenic Living*, held annually at the I.B.S. Correspondence Centers in Costa Rica, Mexico, Canada, and other locations.

Please send me the following books:

____*The Essene Gospel of Peace, Book One $1.00*
____*The Essene Gospel of Peace, Book Two $5.80*
____*The Essene Gospel of Peace, Book Three $5.60*
____*The Essene Gospel of Peace, Book Four $4.50*
____*The Essene Way—Biogenic Living $8.80*
____*The Biogenic Revolution $9.50*
____*The Chemistry of Youth $7.50*
____*From Enoch to the Dead Sea Scrolls $4.80*
____*The Essene Book of Creation $4.50*
____*The Essene Code of Life $3.50*
____*The Essene Science of Life $3.50*
____*Essene Communions with the Infinite $3.95*
____*The Essene Book of Asha $7.50*
____*Archeosophy, a New Science $4.80*
____*Books, Our Eternal Companions $3.50*
____*Cosmos, Man and Society $5.80*
____*Search for the Ageless, Volume One $7.80*
____*The Ecological Health Garden, the Book of Survival $4.50*
____*The Art of Study: the Sorbonne Method $3.50*
_______*Total*
_______*Add 15% for Postage & Handling*

*U.S. $*_______*Total Amount Enclosed*

Please address all orders and correspondence to:

I.B.S. INTERNACIONAL
Box 205, Matsqui, British Columbia, Canada V0X 1S0

*Name*____

*Address*____

*City, State/Province, Zip/Code*____________________________

All orders must be pre-paid. Minimum order: $5.00, minimum postage: $1.00. Please make check, draft or money order in U.S. currency, drawn on a U.S. bank, out to *I.B.S. Internacional.* Allow four weeks for processing. Members of the International Biogenic Society may deduct 20% and *active* Teachers of Biogenic Living or Essene Ministers may deduct 40% (both must have valid memberships for the current year). Dealers and distributors, please write for discount information. All sales are considered final.